Understanding the Elements of the Periodic Table™

THE TRANSACTINIDES

Rutherfordium, Dubnium, Seaborgium, Bohrium, Hassium, Meitnerium, Darmstadtium, Roentgenium

Linley Erin Hall

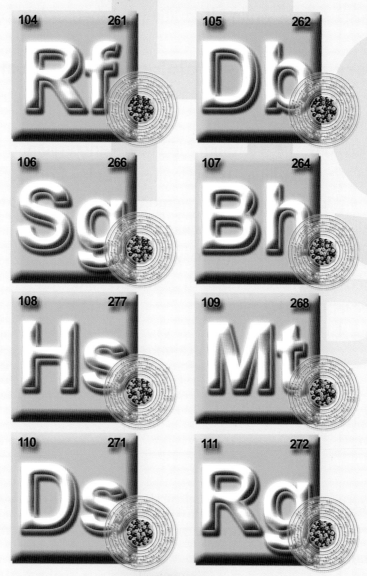

rosen publishing's
**rosen
central**®

New York

Published in 2010 by The Rosen Publishing Group, Inc.
29 East 21st Street, New York, NY 10010

Library of Congress Cataloging-in-Publication Data

Hall, Linley Erin.
The transactinides: rutherfordium, dubnium, seaborgium, bohrium, hassium, meitnerium, darmstadtium, roentgenium / Linley Erin Hall.—1st ed.
    p. cm.—(Understanding the elements of the periodic table)
Includes bibliographical references and index.
ISBN 978-1-4358-3559-7 (library binding)
1. Superheavy elements—Popular works. I. Title.
QD172.S93H35 2010
546'.449—dc22

2009014943

Manufactured in Malaysia
CPSIA Compliance Information: Batch #TW10YA: For Further Information contact Rosen Publishing, New York,
New York at 1-800-237-9932

On the cover: The cover graphic shows the eight known transactinide elements, as they appear on the periodic table. The atomic structure of each element is shown.

Contents

Introduction

In 1869, Dmitry Mendeleyev proposed what he called a periodic table. This chart organized the elements according to their properties. Mendeleyev usually placed elements on the table in order of increasing atomic weight. The elements were also grouped by properties they had in common. But this periodic table contained gaps. Mendeleyev predicted that scientists would discover elements with particular properties to fill those gaps. Over the next two decades, researchers found gallium, scandium, and germanium, which exactly fit gaps in Mendeleyev's table.

Over the years, the periodic table has changed as scientists have learned more about the elements. The table is now organized not by atomic weight but by atomic number. Atomic number is the number of protons (positively charged particles) in an atom of an element. The periodic table also contains many more elements than were known in Mendeleyev's day.

Eventually, scientists discovered all the elements that could be found in nature. So some researchers decided to try to make new elements with larger atomic numbers. They wondered how big an atom could be. They wanted to know if heavier elements would have properties similar to lighter elements on the periodic table. New elements might also have characteristics that would make them useful to humans.

Dmitry Mendeleyev created the first periodic table 140 years ago. The transactinides are the newest additions to the table.

The researchers were successful in making many new elements, although not as successful as many hoped. Still, their work has extended the periodic table. The most recently created elements are the transactinides. The transactinides have atomic numbers of 104 and higher. As far as scientists know, none of the transactinides occur in nature. Researchers have made them all in the laboratory, sometimes with great difficulty. Experiments are occurring even now to create transactinides with higher atomic numbers.

Chemist Paul Karol has calculated that if the historic rate of discovery of new elements (roughly one every two-and-a-half years) continues, the periodic table will include 150 elements by the end of the twenty-first century. Many scientists believe, however, that really heavy elements cannot exist. Only time will tell what the maximum size of the periodic table will be.

Because the transactinides are so new and unstable, little is known about them compared to other elements. But discoveries are made all the time. Research on the transactinides has pushed scientists to create more advanced laboratory equipment that has been used in other experiments as well. This research has also provided new information about atomic structure and properties.

Chapter One
The Transactinides

The transactinides are the elements with atomic numbers of 104 and higher. Their atoms are the largest and heaviest of all the elements. Because of this, the transactinides are sometimes called the superheavy elements. The transactinides that have been named are:

Rutherfordium (Rf): Atomic number 104
Dubnium (Db): Atomic number 105
Seaborgium (Sg): Atomic number 106
Bohrium (Bh): Atomic number 107
Hassium (Hs): Atomic number 108
Meitnerium (Mt): Atomic number 109
Darmstadtium (Ds): Atomic number 110
Roentgenium (Rg): Atomic number 111

Transactinides on the Periodic Table

Most periodic tables have a main block, with two additional rows underneath for the lanthanide and actinide elements. The transactinides that have been discovered so far are found in the last row of the main block of the periodic table. The rows of the table are known as periods. These are numbered from top to bottom.

H																	He
Li	Be											B	C	N	O	F	Ne
Na	Mg											Al	Si	P	S	Cl	Ar
K	Ca	Sc	Ti	V	Cr	Mn	Fe	Co	Ni	Cu	Zn	Ga	Ge	As	Se	Br	Kr
Rb	Sr	Y	Zr	Nb	Mo	Tc	Ru	Rh	Pd	Ag	Cd	In	Sn	Sb	Te	I	Xe
Cs	Ba	La	Hf	Ta	W	Re	Os	Ir	Pt	Au	Hg	Tl	Pb	Bi	Po	At	Rn
Fr	Ra	Ac	Rf	Db	Sg	Bh	Hs	Mt	Ds	Rg							

Ce	Pr	Nd	Pm	Sm	Eu	Gd	Tb	Dy	Ho	Er	Tm	Yb	Lu
Th	Pa	U	Np	Pu	Am	Cm	Bk	Cf	Es	Fm	Md	No	Lr

The modern periodic table contains 111 elements arranged according to their atomic number and properties. The transactinides, or superheavy elements, are found in period (row) 7 and are part of the transition elements.

The transactinides are part of period 7. Within a period, atomic number increases from left to right. There is a jump in atomic numbers in period 7 from 88 to 103 because the actinides listed separately below the table are part of period 7. But the "trans" in "transactinides" indicates that these elements have higher atomic numbers than the actinides. Many periodic tables list an element's atomic number and atomic weight (expressed as atomic mass units, or amu). The atomic weights for the transactinides are usually given as whole numbers in parentheses. This indicates that the atomic weights are estimates.

The periodic table can be divided vertically as well as horizontally. The columns of the main block of the table are known as groups. These are numbered from left to right. For example, rutherfordium is a member of group 4, dubnium of group 5, and so on. Elements in a group usually have some similar properties. Because of this, elements in the same column or group are often referred to as homologues. Rutherfordium's homologues in group 4 are titanium, zirconium, and hafnium.

The middle of the main block of the periodic table, groups 3 to 12, are known as the transition elements. All the named transactinides are transition elements. All transition elements are metals. Most are solid at room temperature (mercury is an exception). Unfortunately, because the transactinides have been created in the laboratory only relatively recently (within the past fifty years), researchers don't yet know much about their properties. Chapters 4 and 5 discuss this in more depth.

Some Transactinide History

All the transactinides have been made by scientists in laboratories. Uranium, element 92, is the heaviest element found in nature. All elements, from atomic number 93 on, have been artificially made. Scientists have created most of them in laboratories. Others were first found in the waste from nuclear explosions.

As scientists have learned more about atoms and atomic structure, ideas about the transactinides have changed. In the late 1930s, Lise Meitner and Otto Frisch predicted that elements with atomic numbers larger than 100 couldn't exist. Using what was known about atoms at the time, they calculated that superheavy elements would fall apart immediately. In the 1950s and 1960s, researchers with new information predicted that elements up to atomic number 164 could exist. They thought that some of these superheavy elements would not be stable at all, but that others would be extremely stable. Scientific cartoons from this period show scientists venturing across a "sea of instability" toward "superheavy island."

An engineer removes a sample of moon rock from a detector in a cramped mini-lab. No transactinides were found in the sample. This provided another piece of evidence that transactinides are not found in nature.

Some scientists believed that some of the transactinides would be stable enough to be present in nature. Researchers looked for transactinides in many different ores. They also examined meteorites and rocks from the moon. In the end, no one found transactinides in nature. These elements turned out to be much more unstable than people had thought. But they can and do exist—in laboratories.

Creating Transactinides

Scientists first began trying to create transactinide elements in the laboratory in the 1960s. Rutherfordium, dubnium, and seaborgium were first

German science and education minister Annette Shavan celebrates the naming of element 111, roentgenium, at the Gesellschaft für Schwerionenforschung (GSI), the laboratory where roentgenium was discovered in the 1990s.

discovered in the 1960s and 1970s. Bohrium, hassium, and meitnerium were discovered in the 1970s and 1980s, and darmstadtium and roentgenium in the mid-1990s.

Creating these elements turned out to be more difficult than researchers expected. The transactinides are not so unstable that they can't exist, but they are still quite unstable. Researchers had to try many different experiments to find ones that worked. They developed new equipment that was more sensitive. Still, experiments often lasted days or weeks before a transactinide was successfully created.

Glenn Seaborg

Element 106, seaborgium, is named for the American chemist Glenn Seaborg (1912–1999), who codiscovered ten elements. In 1951, Seaborg received the Nobel Prize in Chemistry along with Edwin M. McMillan for the actinide hypothesis. Before 1940, the known actinides were placed on the periodic table where the transactinides are today. Seaborg used the chemical properties of elements 93 to 96 to suggest that elements heavier than actinium formed a series similar to the lanthanides. He predicted that

Glenn Seaborg used this ion-exchanger illusion column to discover some of the actinide elements.

elements with atomic numbers greater than 104 would return to the main block of the periodic table. Research since then has confirmed Seaborg's hypothesis.

Scientists are now working to create transactinides with even higher atomic numbers. Some evidence suggests that transactinides with atomic numbers of 112 through 116, as well as 118, have been discovered. Researchers are working to confirm these discoveries. When and if there's enough evidence, these elements will receive permanent names and symbols. Researchers have also tried to create elements 119 and 120 but have not yet reported any success.

Many scientists now think that elements beyond atomic number 126 cannot exist, but that was once mistakenly said about elements with an atomic number higher than 100. New lab techniques, equipment, and technology may someday prove current thinking wrong as well. In any case, it will be exciting to see how the search for these superheavy elements progresses.

Chapter Two
Atomic Structure

The transactinides, like all atoms, are composed of protons, neutrons, and electrons. Protons are positively charged particles. Electrons are negatively charged particles. Neutrons do not have an electric charge. Protons and neutrons make up the center of the atom, known as the nucleus. Electrons form a cloud around this nucleus.

As mentioned, an atom's atomic number is the number of protons that it contains. Changing the number of protons changes the element. In a neutral atom, the number of protons and electrons is the same.

Isotopes

Because neutrons are not charged, atoms of an element can vary in how many they have. Two atoms with the same number of protons, but different numbers of neutrons, are called isotopes. For example, bohrium has ten known isotopes with the numbers of neutrons ranging from 153 to 165. Scientists indicate the isotope of an element in two ways. Both rely on the mass number, which is the sum of an element's protons and neutrons. Bohrium has 107 protons. If an isotope of bohrium has 163 neutrons, the mass number is 107 + 163 = 270. To indicate the isotope, the mass number can be written after the element's

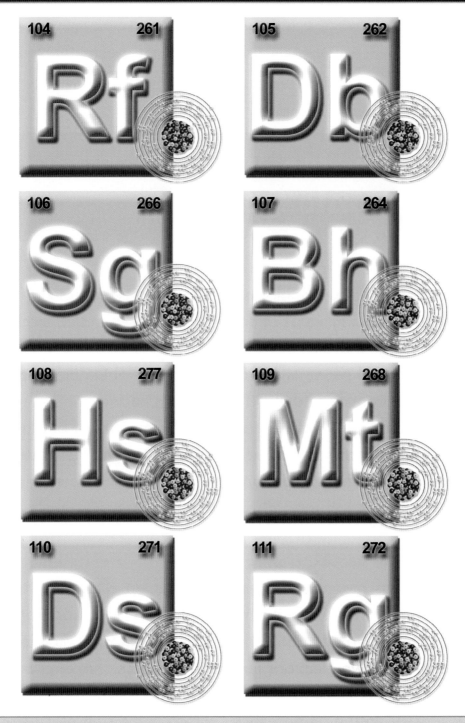

104	261	Rf
105	262	Db
106	266	Sg
107	264	Bh
108	277	Hs
109	268	Mt
110	271	Ds
111	272	Rg

The transactinides are the largest and heaviest elements on the periodic table. Their atomic numbers are in the top left corners, and their estimated atomic masses are in the top right corners. Exact atomic masses are not known because these elements are so unstable.

name: bohrium-270. Or it can be placed to the left of the element's symbol as a superscript: ^{270}Bh.

Atoms with small atomic numbers, like carbon (atomic number 6), usually have numbers of neutrons that are the same as or slightly higher than the number of protons. For example, the three isotopes of carbon have six, seven, and eight neutrons. However, as the atomic number gets larger, the number of neutrons is much higher than that of protons. Rutherfordium-267 has 104 protons and 163 neutrons, for example. These extra neutrons help stabilize the atoms. Although an atom is neutral overall, its nucleus is positively charged. Positive charges repel each other. Neutrons provide padding between positively charged protons and help prevent the nucleus from flying apart.

Electron Shells

The electrons in an atom are constantly moving, forming a negatively charged cloud around the nucleus. At the same time, the electrons are arranged in the atom into specific shells and subshells. These are the areas where electrons can be located based on their energies. The shells and subshells are also one reason why the periodic table has its odd shape.

Shells are numbered, while subshells are known by letters. The lowest shell, 1, has one subshell, s, which can contain up to two electrons. Shell 2 contains one s subshell and one p subshell. The p subshell can contain six electrons. With two electrons in the s subshell and six electrons in the p subshell, shell 2 can contain up to eight electrons total. Shell 3 adds a d subshell that can hold 10 electrons, and shell 4 adds an f subshell that can hold 14 electrons.

Filled shells are known as closed, while partly filled shells are open. Elements in which the highest shell is filled tend to be very stable. The

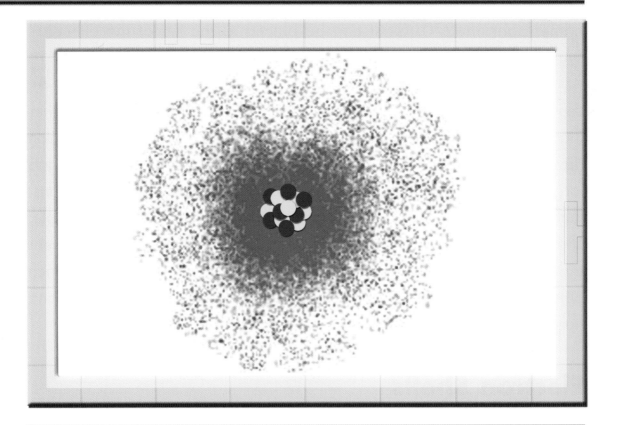

A cloud of electrons (blue dots) surrounds the nucleus of an atom. The density of the cloud shows that more electrons are more likely to be found closer to the nucleus than farther away.

transactinides are large, heavy elements containing a lot of electrons, so they have a lot of filled shells. But their highest shell is not filled. The transition elements, of which the transactinides are part, are also known as the d-block of the periodic table. This is because a d subshell is being filled in these elements. The electrons in this d subshell are available to react with other elements. Chapter 5 talks about what researchers have learned about compounds made with transactinides.

The electrons in the highest partly filled shell are available for chemical reactions. An atom that has lost or gained electrons from a neutral atom is called an ion. Positive ions have lost electrons, which leaves them

Niels Bohr

Element 107, bohrium, is named after Danish physicist Niels Bohr (1885–1962). Bohr developed the concept of electron shells. He suggested that each electron orbits at a specific distance from the nucleus, depending on its energy. This distance is determined by the electron's shell and subshell, which are filled in a particular order. In the periodic table, elements in the same group

Niels Bohr made many important discoveries about atomic structure that helped scientists learn why certain elements have similar characteristics.

have similar properties because they have the same number of electrons in their outer shells. Bohr received the Nobel Prize in Physics in 1922.

with an overall positive charge. Negative ions have gained electrons, giving them an overall negative charge. Each element tends to form ions with certain charges. Ions in which an outer subshell is either completely full or completely empty are less reactive. Transition metals like the transactinides are often found as several different positive ions. The research that has been done on transactinide ions so far has only turned up one ion for each transactinide, however.

Proton and Neutron Shells

Scientists have also found that neutrons and protons also form shells in the nucleus. Like with electron shells, a closed proton and/or neutron shell makes a nucleus more stable. The proton and neutron numbers at which shells are closed are often called magic numbers. The known magic numbers are 2, 8, 20, 28, 50, and 82. Neutrons also have a closed shell at 126. Scientists have predicted that 114, 120, and 126 might also be magic numbers for protons, and that 184 might be a magic number for neutrons.

Because of the extra stability corresponding to the magic numbers, scientists tried to create some heavier transactinide elements before lighter ones. Magic numbers also help explain why some researchers thought that some super-stable transactinides existed on "superheavy island," while other, lighter ones were lost in the "sea of instability."

No transactinides are truly stable, however. The next chapter discusses the reasons for this instability.

Chapter Three
Radioactivity

Isotopes can be stable or radioactive. A stable atom is one that will not change into an atom of another element through radioactive decay. The atoms that make up the things people use in their everyday lives are almost all stable. On the other hand, a radioactive element is unstable. Through one of several different processes, it will eventually change into another element. Usually, this transformation occurs by the atom emitting energy and/or a particle. These processes are collectively known as radioactive decay.

Most elements have some isotopes that are stable and others that are radioactive. Chapter 2 gave carbon as an example of an element with three isotopes. The

An atom of element 112 emits a helium nucleus, also called an alpha particle, to become an atom of element 110, darmstadtium.

carbon isotopes with six and seven neutrons are stable, but the carbon isotope with eight neutrons is radioactive. However, all the isotopes of all the transactinide elements are radioactive. In fact, all isotopes of all elements with atomic numbers greater than 82 are radioactive.

Half-Lives

The half-life of a radioactive isotope is the average length of time needed for half of a quantity of that isotope to decay into another element. Say you have 128 atoms of bohrium-270, which has a half-life of about one minute. After a minute, you'll have sixty-four atoms of

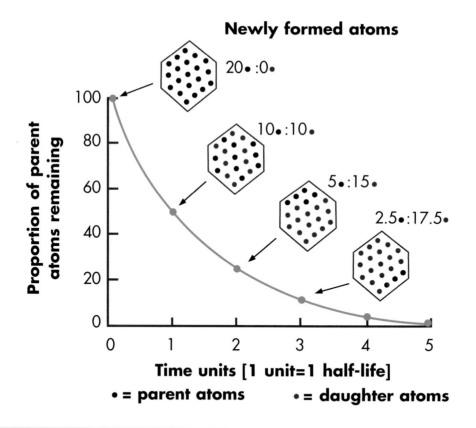

These parent atoms have a half-life of one time unit. At time zero, there are twenty of them. After one time unit, ten, or half, have decayed into daughter atoms. This decay continues until all parent atoms have decayed.

bohrium, on average. After two minutes, you'll have thirty-two atoms, on average, and so on, until all the bohrium atoms have decayed. However, all the atoms don't decay precisely at the one-minute mark. They can decay at any point during that minute-long period.

Different isotopes of an element can have very different half-lives. For example, dubnium-260 has a half-life of 1.6 seconds, but dubnium-268 has a half-life of twenty-eight hours. This is the longest half-life of any of the transactinides. Most isotopes of transactinide elements have half-lives of seconds or fractions of seconds, like dubnium-260. These short half-lives help explain why so little is known about the transactinides. It's hard to perform experiments on an atom that only exists for a few seconds or less. Researchers have had to create new equipment to detect and measure elements that exist for such brief periods of time.

These short half-lives also help explain why the transactinides are not found in nature. The radioactive isotope of carbon, carbon-14, has a half-life of 5,730 years. It exists long enough that scientists can detect it in living things. But even if the transactinides did exist in nature at some point in the past, they decayed long ago and can no longer be found.

Radioactivity: Dangerous Yet Useful

Radioactivity is dangerous. The highly energetic particles produced during radioactive decay can harm cells and cause many health problems. Researchers who work with radioactive elements take safety precautions. However, in small quantities, radioactive materials have been used to improve human health. For example, several methods that doctors use to see inside the body without resorting to surgery rely on radioactive tracers. Radioactive isotopes are also used to attack, shrink, and kill cancerous tumors in the body.

The transactinides are produced one atom at a time. This makes it impossible to measure when half of the element is gone. Researchers instead measure how long it takes that particular atom to decay. This is known as the lifetime. The half-life is usually 30 percent shorter than the lifetime. As scientists gather more data, they are more accurately able to measure both lifetime and half-life for newly created elements.

Radioactive Decay and Fission

There are several types of radioactive decay. The two types that transactinides are most likely to undergo are alpha decay and electron capture.

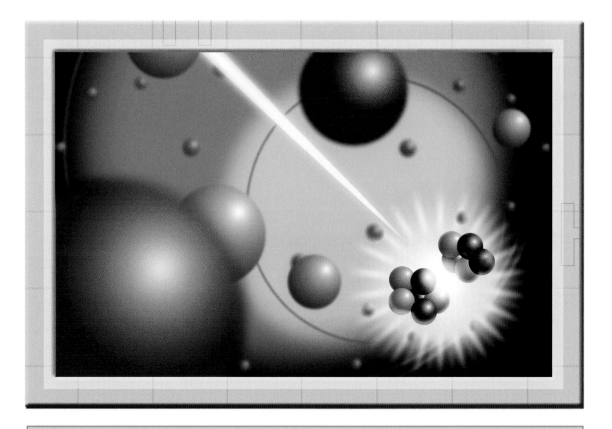

In nuclear fission, an atom splits into two smaller atoms. This process can release a vast amount of energy. Indeed, it's how nuclear power plants produce energy. Some isotopes of transactinides also decay via fission.

The transactinides may also split apart due to spontaneous fission. Some isotopes always decay by the same method. Other isotopes can decay by multiple methods.

In alpha decay, an atom emits the nucleus of a helium atom. The nucleus, which contains two neutrons and two protons, is often called an alpha particle. When an element undergoes alpha decay, it transforms into the element with an atomic number two less than it started with. For example, hassium-265, with an atomic number of 108, undergoes alpha decay to become seaborgium-261, with an atomic number of 106.

In electron capture, a proton captures one of the electrons in the cloud around the nucleus of the atom. The proton and electron combine to form

Lise Meitner

Element 109, meitnerium, is named for Austrian physicist Lise Meitner (1878–1968), a codiscoverer of fission. Although she was forced to flee Nazi Germany because she was Jewish, Meitner helped design the experiments that her colleagues Otto Hahn and Fritz Strassman later conducted to prove the existence of fission. She also wrote up the physical explanation of the experiment. However, since Meitner was a woman and was not present for the experiments themselves, many ignored her contributions.

Lise Meitner was excluded from the Nobel Prize her colleague Otto Hahn won for the discovery of fission. She is the only one with an element named for her, however.

a neutron and a particle called a neutrino. The atom emits the neutrino, but the neutron stays in the nucleus. Electron capture decreases an atom's atomic number by one and increases its neutron number by one. For example, dubnium-258 (105 protons, 153 neutrons) can decay by electron capture to rutherfordium-258 (104 protons, 154 neutrons).

Many transactinide isotopes also undergo spontaneous fission. Fission is the splitting of one atom into two new atoms with smaller atomic numbers. These new atoms usually have similar atomic masses. Large atoms like the transactinides are more likely to undergo spontaneous fission. Scientists can also make some atoms undergo fission by bombarding them with small particles, such as neutrons.

After an element undergoes radioactive decay, the isotope of the new element formed may be stable or radioactive. If it is radioactive, it will also decay according to its half-life. In many cases, researchers can follow a decay chain of different elements until a stable isotope is reached. The next chapter will look at how scientists used decay chains to discover the transactinides.

Chapter Four
Creating the Transactinides

Three major research groups have worked on creating the transactinide elements in their laboratories. One group is centered at the Lawrence Berkeley National Laboratory in Berkeley, California. A second is based at the Gesellschaft für Schwerionenforschung (GSI), in Darmstadt, Germany. The third group is based at the Joint Institute for Nuclear Research in Dubna, Russia. When research on the transactinides began, Russia was part of the Union of Soviet Socialist Republics (USSR). The United States and USSR were in a Cold War, political and military tensions were high, and cooperation between the two nations was ordinarily low.

This linear accelerator at the GSI in Germany was used for experiments that discovered several transactinides. Its total length is 394 feet (120 meters).

Despite this, the three groups competed and cooperated with each other. Each lab wanted to be the first to create a new element, but they also visited each other's facilities and ran some joint experiments. When one lab announced a discovery, the others would try to replicate the experiment to verify the claim. Some disagreements arose as to who first synthesized certain transactinide elements, particularly 104 and 105. Chapter 6 talks about this controversy and how it affected the naming of the transactinides.

Smashing Atoms

To synthesize the transactinide elements, scientists use fusion. Fusion is the opposite of fission. It's the combination of two atoms into one atom that has a larger atomic number. This combination of atoms requires a lot of energy. To give the atoms the energy they need to fuse, atoms of one element are accelerated to a very high speed in a particle accelerator. Then a beam of these atomic projectiles is directed at a target made up of atoms of the second element. Usually, the lighter element forms the projectile, and the heavier element is the target. Particle accelerators are sometimes called atom smashers for good reason.

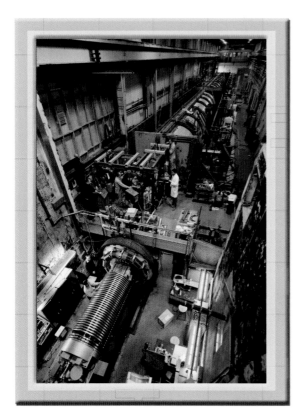

This linear accelerator at the Lawrence Berkeley National Laboratory was used to create transactinides for many experiments.

Powerful electric or magnetic fields are used to accelerate atoms in accelerators. An atomic accelerator can be either linear or circular. The German group used a linear accelerator, while the Russian and American groups worked on circular accelerators. Over the years, the three groups of researchers upgraded their equipment so that they could accelerate heavier atoms and safely handle radioactive beams. This improved their ability to create transactinides.

Researchers choose the elements in the beam and the target based on what element they are trying to create. It's important to use isotopes that are rich in neutrons, since extra neutrons help stabilize superheavy elements. Because the different laboratories had different equipment, the researchers sometimes used different elements in their beams and targets to produce the same transactinide. In some cases, this meant that two groups would synthesize different isotopes of the same element within months of each other.

For example, in their synthesis of dubnium, the American group reacted carbon-12 with californium-249 to produce dubnium-257 (plus four neutrons). The Russian group, on the other hand, used neon-22 with plutonium-242 to produce dubnium-259 (plus five neutrons).

Making Fusion Happen

Many things need to happen just right in order for fusion to occur. First, the nuclei have to collide. Most of an atom is the electron cloud, and the nucleus is very small in comparison. Therefore, collisions between the nuclei of two atoms are extremely rare. Second, the two atoms must collide head-on. If two nuclei simply graze each other, fusion will not occur. Third, the projectile must be traveling at a particular speed and thus have a certain energy. If the energy is too little, fusion will not occur. The atoms will just bounce off each other instead. If the energy is too high, the elements will immediately fission. According to Sigurd Hoffman in his book

Deuterium-Tritium Fusion Reaction

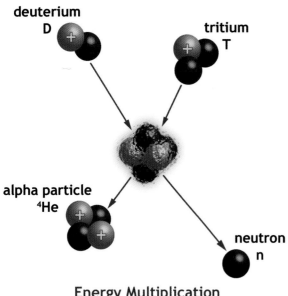

deuterium
D

tritium
T

alpha particle
^{4}He

neutron
n

Energy Multiplication
about 450:1

> In a fusion reaction, two atoms collide and combine to create a single new atom with a higher atomic number. The transactinides are made using fusion.

On Beyond Uranium, the minimum projectile speed needed to produce an element with an atomic number of 100 to 110 is 30,000 km/second (18,641 miles/second), about 10 percent the speed of light. However, even if the energy is just right, fusion may still not occur.

Because of these difficulties, researchers often need to run experiments for a very long time. For example, the original experiment to create bohrium-262 ran twenty-four hours a day for six days. During this entire time, the researchers detected only six atoms of the new element (meaning there were only six successful fusions). These atoms were produced and detected one at a time.

Detecting Atoms

After an atom of a transactinide is produced, it must be detected and identified. After fusion occurs, the new superheavy element moves in the direction of the projectile beam. The transactinide must be separated from the atoms of the beam in order to be identified. This is particularly true if the target or projectile is radioactive. Aerosol gas jets are often used to separate the various atoms based on weight.

Ernest Rutherford

Element 104, rutherfordium, is named after Ernest Rutherford (1871–1937), a physicist from New Zealand. Two of Rutherford's discoveries are important to the synthesis of superheavy elements. First, he identified two types of radioactive decay, which he named alpha and beta. Rutherford won the Nobel Prize in Chemistry for this work in 1908. Second, Rutherford discovered that an atom's positive charge and most of its mass are concentrated in a tiny nucleus.

Ernest Rutherford is sometimes called "the father of nuclear physics" because of all the important discoveries he made concerning atoms.

Transactinide elements are usually identified based on their decay patterns. The first isotope in a decay chain is called a parent. The later isotopes are called daughters. The researchers try to set up experiments in which the daughters of the transactinides are isotopes with known half-lives and decay energies. A decay chain consisting of several members and clear half-lives and energies is good evidence for the discovery of a new transactinide. These decay chains can also lead to the discovery of new isotopes. Many transactinide isotopes were first identified as daughters of elements with higher atomic numbers.

Once scientists have created and identified a new transactinide, they want to learn about its unique properties. The next chapter discusses what scientists have learned and why this information has been hard to obtain.

Chapter Five
Properties and Compounds

After researchers discover a new element, they want to explore its properties. These may include the element's ability to form compounds with other elements. Determining the properties and possible compounds of the transactinides has been very difficult, however. There are two reasons for this.

First, as explored in chapter 3, the transactinides are radioactive and have very short half-lives. Researchers can't take a sample down the hall and run an experiment on it that takes several minutes if the sample decays in mere seconds. Second, as discussed in chapter 4, experiments in accelerators produce very small quantities of the transactinides, sometimes only one atom in an entire day. Performing chemistry on a single atom is not an easy task.

One-Atom-at-a-Time Chemistry

To address these challenges, scientists have developed online, one-atom-at-a-time chemistry. "Online" means the apparatus needed for the chemical reaction is attached to, or part of, the detector in the accelerator. This shortens the transit time so that the experiment can begin more quickly. "One-atom-at-a-time" means the chemistry is optimized for the smallest amount of the element possible: one atom.

However, one-atom-at-a-time chemistry cannot reveal certain things about the transactinides. Based on their placement on the periodic table, a scientist can predict certain characteristics of the transactinides. For example, as transition elements, transactinides should be metals that are solid at room temperature. But their melting points have not been determined experimentally—neither have their boiling points, conductivities, or densities. No one knows what color the transactinides are or how hard they are. Determining these kinds of properties experimentally requires larger samples than just one atom. Using computer models, scientists have calculated possible values for some of these properties. But only time will tell if these calculations are correct.

The experiments that scientists have carried out with the transactinides have focused on chemical reactions with other elements and compounds. Experiments with fluorine, chlorine, and bromine are the most popular. Some transactinides have also been observed to form compounds with oxygen. Most of these experiments have been carried out with the transactinides as gases. Right after fusion, atoms are extremely hot. Thus, working in the gas phase makes sense. However, some chemistry has been done with the transactinides dissolved in water.

Using Homologues

As mentioned in chapter 1, elements in the same group of the periodic table, called homologues, usually have similar properties. Scientists have used the homologues of the transactinides to design their experiments. For example, they prepare for experiments with the transactinides by running those same experiments on their homologues. This allows scientists to figure out the best methods and conditions for their transactinide experiments.

One technique that researchers use is to create a target that contains a bit of an impurity. The projectile can fuse with this impurity to

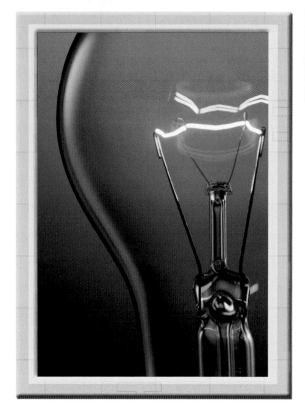

Tungsten, which is used in the filaments of lightbulbs, is a homologue of seaborgium. They are expected to have similar properties.

create an isotope of a transactinide homologue. For example, consider a target made of curium and gadolinium. The fusion of neon-22 and curium-248 yields seaborgium-265. Similarly, the fusion of neon-22 and gadolinium-152 yields tungsten-168, a homologue of seaborgium.

Mixing a transactinide and a homologue allows researchers to compare their behavior under the same experimental conditions. This also helps researchers verify that the reaction is working as planned. Furthermore, if a transactinide's properties are similar to that of its homologues, it's more likely that the transactinide has been properly placed on the periodic table.

Transactinide Compounds

Scientists know more about rutherfordium and dubnium than most other transactinides. They were the first transactinides to be discovered and have been studied the longest. They are also produced in higher quantities than the other transactinides—they are created at rates of atoms per hour, rather than atoms per day. However, researchers still don't know a lot about even these relatively accessible transactinides.

Hafnium, a silvery, lustrous metal, is a homologue of rutherfordium. Since the transactinides are produced one atom at a time, no one knows what they look like, but scientists think that rutherfordium probably looks similar to hafnium.

Rutherfordium is a member of group 4 of the periodic table. Its homologues are titanium, zirconium, and hafnium. Researchers have found that rutherfordium forms a Rf^{4+} ion and the compounds $RfCl_4$, $RfBr_4$, $RfOCl_2$, and $K_2[RfCl_6]$. Its homologues form similar compounds. Dubnium is a member of group 5, with homologues of vanadium, niobium, and tantalum. Four dubnium compounds have been synthesized: $DbCl_5$, $DbBr_5$, $DbOCl_3$, and $DbOBr_3$.

Seaborgium was discovered a few years after rutherfordium and dubnium. For more than twenty years, however, seaborgium-263 was the longest-lived transactinide isotope known. It has a half-life of only

Heavy Atoms

The transactinides are much heavier than their homologues. For example, seaborgium-265 is more than 50 percent heavier than its homologue tungsten-168. Transactinides have big, heavy nuclei with lots of protons and neutrons. Nuclei like this pull their electrons in closer than lighter elements do. This can cause them to behave in different ways than their homologues do. However, the transactinides thus far have been found to be similar enough to their homologues that no one questions their placement on the periodic table.

0.9 seconds, too short for chemistry. In the 1990s, scientists finally created seaborgium isotopes with longer half-lives that they could perform chemistry on. Still, relatively little is known about this element. Research has shown that seaborgium forms a Sg^{6+} ion and the compounds SgO_2Cl_2, SgO_2F_2, SgO_3, and $SgO_2(OH)_2$. It's a member of group 6 and has chromium, molybdenum, and tungsten as homologues.

Bohrium is a member of group 7, with homologues of manganese, technetium, and rhenium. Only one bohrium compound has been synthesized, BhO_3Cl. Hassium is a member of group 8, with homologues of iron, ruthenium, and osmium. Two compounds of hassium have been made, HsO_4 and $Na_2[HsO_4(OH)_2]$. Although there isn't a lot of evidence to go on, the similarities between these compounds and those of the homologues of bohrium and hassium were enough to convince scientists that they were in the proper place on the periodic table.

No compounds have been created yet with elements 109 to 111 because their half-lives are just too short. However, scientists continue to search for new isotopes of all transactinides. New discoveries about these elements' properties and compounds may be announced at any time.

Chapter Six
Naming the Transactinides

Humans have used elements like copper, tin, and gold for thousands of years. When scientists began to classify the elements, they usually kept the names that people already used for them. They've also discovered many previously unknown elements, including the transactinides. It is important for the scientific community to agree on names for these new elements. In the case of the transactinides, agreement was hard to achieve.

The Controversy

Traditionally, the discoverer of an element has the right to name it. Over the years, the discoverers of

This copper coin is nearly 1,500 years old. When scientists discovered that copper is an element, it already had a commonly used name. The newly discovered transactinides needed to be named, however.

many elements have been obvious and uncontested. But the German, Russian, and American laboratories working on the transactinides often had experiments running at the same time to discover particular elements. When the results of an experiment indicate the discovery of a new element, other researchers evaluate the claim. Sometimes, opinions differ.

The first reports of the synthesis of element 104 came from the group of scientists in Dubna, Russia. Many researchers found these experiments to be inconclusive. The American scientists at the Lawrence Berkeley National Laboratory were working on element 104 at the same time. Their first results were published after those from Dubna, but many people thought the Berkeley group's claim to creating element 104 was stronger than that of the Russians. A similar situation existed for element 105.

In 1974, the International Union of Pure and Applied Chemistry (IUPAC) formed a committee to resolve the dispute about the discovery of these elements. The IUPAC is the organization that makes the rules about how to name elements, compounds, and chemical reactions. The committee facilitated meetings between the American and Russian groups, but no conclusions were reached.

For many years, Americans referred to element 105 as hahnium after Otto Hahn. He codiscovered fission with Lise Meitner.

Meanwhile, in America and Germany, element 104 was named rutherfordium, and element 105 was named hahnium (after Otto Hahn, codiscoverer of fission with Lise Meitner). In Russia, however, element 104 was named kurchatovium, after Igor Kurchatov, who developed atomic weapons in the Soviet Union. They named element 105 nielsbohrium. The controversy over naming these elements held up the naming of other transactinides.

In 1986, the International Union of Pure and Applied Physics (IUPAP) decided to set up its own committee to review the situation. A couple of scientists from the IUPAC served on this committee, which published reports in 1991 and 1992. It gave shared credit to the Berkeley and Dubna groups for elements 104 and 105. This angered many of the researchers involved in the discoveries and did not help establish names for the elements.

In 1994, using the committee's report, the IUPAC developed a list of suggested names for elements 104 to 109. Very few scientists were satisfied with the list. The IUPAC came up with another list in 1995, and, in 1997, the names now in use were finally chosen.

The Names Eventually Agreed Upon

The IUPAC suggests five sources for the names of newly discovered elements: a mythological concept or character (including astronomical objects), a mineral, a place, a property of the element, or a scientist. As noted in the sidebars throughout this book, many of the transactinides are named after scientists who made contributions to our understanding of atoms, elements, and radioactivity. These include Ernest Rutherford (rutherfordium), Glenn Seaborg (seaborgium), Niels Bohr (bohrium), Lise Meitner (meitnerium), and Wilhelm Conrad Roentgen (roentgenium).

Naming element 106 after Glenn Seaborg was controversial, not because he didn't deserve it, but because he was still alive at the time.

Element 111 is named after Wilhelm Conrad Roentgen, who discovered X-rays. Many transactinides are named after scientists.

Some researchers did not feel it was appropriate to name an element after a living scientist. Because his contributions to the discovery of new elements were so many and so important, however, the objections were overruled. Seaborgium is the only element to be named after a living person.

The names of the other transactinides come from the locations of the laboratories where researchers investigated them. Element 105 was named dubnium, after Dubna, Russia. Element 110 is darmstadtium, after Darmstadt, Germany. The name hassium for element 108 also honors the location of the German laboratory. Hassia is the Latin form of Hesse, the German state in which Darmstadt is located. None of the transactinides are named for the Lawrence Berkeley National Laboratory because three actinides already have names associated with its location: americium (America), berkelium (Berkeley), and californium (California). Thus, the American researchers suggested names honoring scientists for the transactinides they made.

Until an element receives a permanent name, it is referred to by a systematic element name. The IUPAC created this system in 1979. The systematic element name system uses the Greek and Latin words for different numbers to create a name based on the element's atomic number. For

Wilhelm Conrad Roentgen

Element 111, roentgenium, was named for Wilhelm Conrad Roentgen (1845–1923), a German scientist who discovered X-rays. X-rays are a form of electromagnetic radiation, like visible light. However, the rays are invisible to the eye and can penetrate many materials that light cannot.

X-rays are commonly used to diagnose problems with the skeleton. Indeed, Roentgen's first publication on X-rays included a photograph of the bones of his wife's hand. He received the Nobel Prize in Physics in 1901.

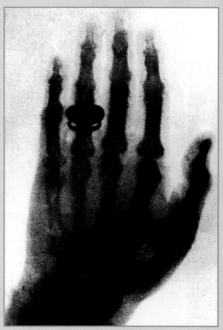

This early X-ray photograph shows the hand of Wilhelm Roentgen's wife. People often have X-rays taken when they go to the dentist or break a bone.

example, element 112 is ununbium, and element 114 is ununquadium. Three letter symbols go along with each of these names. When this system was first introduced, some scientists thought these systematic names should be the permanent names for new elements. Most researchers disliked that idea, though.

After the controversies surrounding the naming of the elements 104 to 109, the IUPAC drew up new guidelines about how to name elements. The names for darmstadtium and roentgenium, both discovered by scientists at GSI (the German lab), were not controversial. It will be interesting to see what scientists choose as the names for elements 112 and higher.

The Periodic Table of Elements

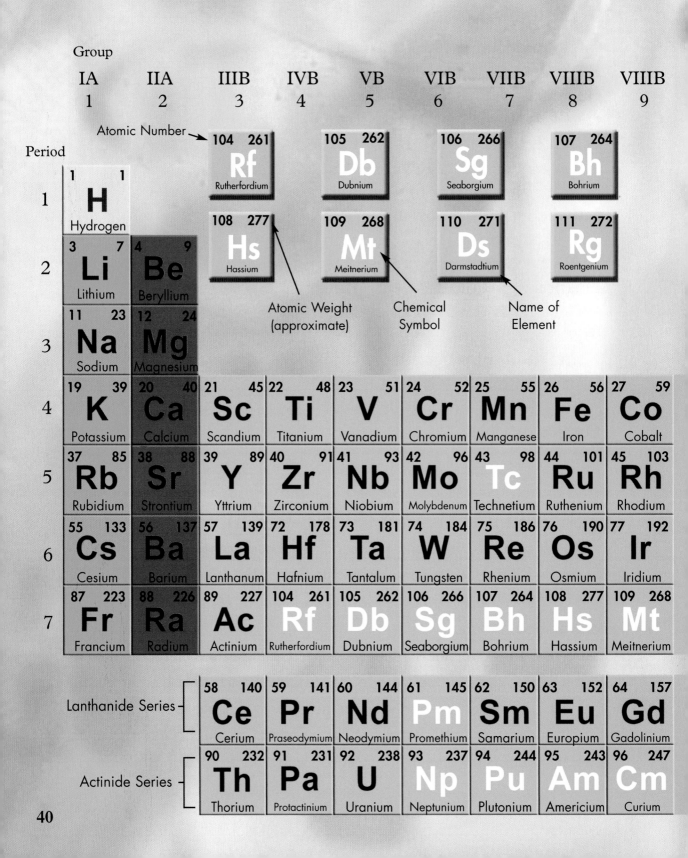

Group

IA	IIA	IIIB	IVB	VB	VIB	VIIB	VIIIB	VIIIB
1	2	3	4	5	6	7	8	9

Atomic Number →

Period

| | 104 261 Rf Rutherfordium | | 105 262 Db Dubnium | | 106 266 Sg Seaborgium | | 107 264 Bh Bohrium | |

| | 108 277 Hs Hassium | | 109 268 Mt Meitnerium | | 110 271 Ds Darmstadtium | | 111 272 Rg Roentgenium | |

Atomic Weight (approximate) Chemical Symbol Name of Element

1 — 1 1 H Hydrogen

2 — 3 7 Li Lithium | 4 9 Be Beryllium

3 — 11 23 Na Sodium | 12 24 Mg Magnesium

4	19 39 K Potassium	20 40 Ca Calcium	21 45 Sc Scandium	22 48 Ti Titanium	23 51 V Vanadium	24 52 Cr Chromium	25 55 Mn Manganese	26 56 Fe Iron	27 59 Co Cobalt
5	37 85 Rb Rubidium	38 88 Sr Strontium	39 89 Y Yttrium	40 91 Zr Zirconium	41 93 Nb Niobium	42 96 Mo Molybdenum	43 98 Tc Technetium	44 101 Ru Ruthenium	45 103 Rh Rhodium
6	55 133 Cs Cesium	56 137 Ba Barium	57 139 La Lanthanum	72 178 Hf Hafnium	73 181 Ta Tantalum	74 184 W Tungsten	75 186 Re Rhenium	76 190 Os Osmium	77 192 Ir Iridium
7	87 223 Fr Francium	88 226 Ra Radium	89 227 Ac Actinium	104 261 Rf Rutherfordium	105 262 Db Dubnium	106 266 Sg Seaborgium	107 264 Bh Bohrium	108 277 Hs Hassium	109 268 Mt Meitnerium

Lanthanide Series

| 58 140 Ce Cerium | 59 141 Pr Praseodymium | 60 144 Nd Neodymium | 61 145 Pm Promethium | 62 150 Sm Samarium | 63 152 Eu Europium | 64 157 Gd Gadolinium |

Actinide Series

| 90 232 Th Thorium | 91 231 Pa Protactinium | 92 238 U Uranium | 93 237 Np Neptunium | 94 244 Pu Plutonium | 95 243 Am Americium | 96 247 Cm Curium |

VIIIB 10	IB 11	IIB 12	IIIA 13	IVA 14	VA 15	VIA 16	VIIA 17	O 18
								2 4 **He** Helium
			5 11 **B** Boron	6 12 **C** Carbon	7 14 **N** Nitrogen	8 16 **O** Oxygen	9 19 **F** Fluorine	10 20 **Ne** Neon
			13 27 **Al** Aluminum	14 28 **Si** Silicon	15 31 **P** Phosphorus	16 32 **S** Sulfur	17 35 **Cl** Chlorine	18 40 **Ar** Argon
28 59 **Ni** Nickel	29 64 **Cu** Copper	30 65 **Zn** Zinc	31 70 **Ga** Gallium	32 73 **Ge** Germanium	33 75 **As** Arsenic	34 79 **Se** Selenium	35 80 **Br** Bromine	36 84 **Kr** Krypton
46 106 **Pd** Palladium	47 108 **Ag** Silver	48 112 **Cd** Cadmium	49 115 **In** Indium	50 119 **Sn** Tin	51 122 **Sb** Antimony	52 128 **Te** Tellurium	53 127 **I** Iodine	54 131 **Xe** Xenon
78 195 **Pt** Platinum	79 197 **Au** Gold	80 201 **Hg** Mercury	81 204 **Tl** Thallium	82 207 **Pb** Lead	83 209 **Bi** Bismuth	84 209 **Po** Polonium	85 210 **At** Astatine	86 222 **Rn** Radon
110 271 **Ds** Darmstadtium	111 272 **Rg** Roentgenium							

65 159 **Tb** Terbium	66 163 **Dy** Dysprosium	67 165 **Ho** Holmium	68 167 **Er** Erbium	69 169 **Tm** Thulium	70 173 **Yb** Ytterbium	71 175 **Lu** Lutetium
97 247 **Bk** Berkelium	98 251 **Cf** Californium	99 252 **Es** Einsteinium	100 257 **Fm** Fermium	101 258 **Md** Mendelevium	102 259 **No** Nobelium	103 262 **Lr** Lawrencium

Glossary

alpha decay A radioactive decay process in which an atom emits two protons and two neutrons.

atom The fundamental unit of an element.

atomic number The number of protons in the nucleus of an atom.

electron A negatively charged particle that is part of an atom.

electron capture A radioactive decay process in which a proton captures one of the electrons orbiting it.

fission The breaking apart of an atom into two atoms with smaller atomic numbers.

fusion The combination of two atoms into a single new atom with a larger atomic number.

half-life The amount of time needed for half a quantity of a radioactive element to decay.

homologues Elements that are part of the same group of the periodic table. They usually have similar properties.

isotopes Two atoms of the same element that contain different numbers of neutrons.

neutron A neutral particle located in the nucleus of an atom.

nucleus The core of an atom, which contains protons and neutrons, has a positive charge, and contains most of the mass of the atom.

particle accelerator A scientific instrument that makes atoms and other particles move very fast.

proton A positively charged particle located in the nucleus of an atom.

radioactive decay The collection of processes by which an atom of one element can change into an atom of a different element.

transactinides Elements with atomic numbers of 104 and higher.

For More Information

American Chemical Society (ACS)
1155 Sixteenth Street NW
Washington, DC 20036
(800) 227-5558
Web site: http://www.acs.org
The ACS is an organization of more than 160,000 chemists and
related scientists—the largest scientific society in the world. It pub-
lishes scientific journals, holds conferences, and tries to educate the
public about chemistry.

Canadian Society for Chemistry (CSC)
130 Slater Street, Suite 550
Ottawa, ON K1P 6E2
Canada
(613) 232-6252
Web site: http://www.cheminst.ca
The CSC is the national technical association representing the field of
chemistry and the interests of chemists in industry, academia, and
government.

International Union of Pure and Applied Chemistry (IUPAC)
P.O. Box 13757
Research Triangle Park, NC 27709-3757
(919) 485-8700
Web site: http://www.iupac.org
The IUPAC addresses global issues related to chemistry. In particular, it
creates the rules for naming elements and compounds.

International Union of Pure and Applied Physics (IUPAP)
American Physical Society
One Physics Ellipse
College Park, MD 20740-3844
(301) 209-3269
Web site: http://www.iupap.org
The IUPAP promotes cooperation between physicists around the world. In particular, it sets standards for the symbols, units, and names used in physics.

Lawrence Berkeley National Laboratory
1 Cyclotron Road
Berkeley, CA 94720
(510) 486-4000
Web site: http://www.lbl.gov
The Lawrence Berkeley National Laboratory is part of the national laboratory system supported by the U.S. Department of Energy. Researchers here have worked on the transactinides for decades, and research continues today.

Web Sites

Due to the changing nature of Internet links, Rosen Publishing has developed an online list of Web sites related to the subject of this book. This site is updated regularly. Please use this link to access this list:

http://www.rosenlinks.com/uept/tran

For Further Reading

Bankston, John. *Lise Meitner and the Atomic Age*. Hockessin, DE: Mitchell Lane Publishers, 2003.

Dingle, Adrian. *The Periodic Table: Elements with Style!* New York, NY: Kingfisher Books, 2007.

Jerome, Kate Boehm. *Atomic Universe: The Quest to Discover Radioactivity*. Des Moines, IA: National Geographic Children's Books, 2006.

Miller, Ron. *The Elements*. Minneapolis, MN: Twenty-First Century Books, 2004.

Newmark, Ann. *Chemistry* (DK Eyewitness Books). New York, NY: DK Children, 2005.

Pasachoff, Naomi. *Ernest Rutherford: Father of Nuclear Science*. Berkeley Heights, NJ: Enslow Publishers, 2005.

Slade, Suzanne. *Elements and the Periodic Table* (Library of Physical Science). New York, NY: PowerKids Press, 2007.

Spangenburg, Ray, and Diane Kit Moser. *Niels Bohr: Atomic Theorist*. New York, NY: Chelsea House Publications, 2008.

Tocci, Salvatore. *The Periodic Table* (A True Book). New York, NY: Children's Press, 2004.

Zannos, Susan. *Dimitri Mendeleyev and the Periodic Table*. Hockessin, DE: Mitchell Lane Publishers, 2004.

Bibliography

American Nuclear Society. "Nuclear Science and Technology." Retrieved February 4, 2009 (http://aboutnuclear.org/view.cgi?fC=NST).

Beckett, M. A., and A. W. G. Platt. *The Periodic Table at a Glance*. Oxford, England: Blackwell Publishing, 2006.

Brock, William H. *The Chemical Tree: A History of Chemistry*. New York, NY: W. W. Norton and Company, 1992.

Hoffman, Darleane C., Albert Ghiorso, and Glenn T. Seaborg. *The Transuranium People: The Inside Story*. London, England: Imperial College Press, 2000.

Hofmann, Sigurd. *On Beyond Uranium: Journey to the End of the Periodic Table*. New York, NY: Taylor and Francis, 2002.

Koppenol, W. H. "Naming of New Elements: IUPAC Recommendations 2002." *Pure and Applied Chemistry*, Vol. 74, No. 5, 2002, pp. 787–791.

Morris, Richard. *The Last Sorcerers: The Path from Alchemy to the Periodic Table*. Washington, DC: Joseph Henry Press, 2003.

Nobel Foundation. *Nobel Lectures, Physics 1901–1921*. New York, NY: Elsevier Publishing Company, 1967.

Rouvray, Dennis H., and R. Bruce King, eds. *The Periodic Table: Into the 21st Century*. Baldock, England: Research Studies Press, 2004.

Schädel, Matthias, ed. *The Chemistry of Superheavy Elements*. Dordrecht, The Netherlands: Kluwer Academic Publishers, 2003.

Seaborg, Glenn T., and Walter D. Loveland. *The Elements Beyond Uranium*. New York, NY: John Wiley & Sons, 1990.

Index

About the Author

Linley Erin Hall is a science writer and editor based in Berkeley, California, where she enjoys hiking in the hills above Lawrence Berkeley National Laboratory, one of the centers of the transactinide elements research. She has a B.S. degree in chemistry from Harvey Mudd College and a certificate in science communication from the University of California, Santa Cruz. The transactinides were not covered in any of Hall's chemistry classes, so she has enjoyed the opportunity to learn about these interesting but often forgotten elements. This is her sixth book for Rosen Publishing. She is also the author of *Who's Afraid of Marie Curie?: The Challenges Facing Women in Science and Technology.*

Photo Credits

Cover, pp. 1, 7, 14, 40–41 by Tahara Anderson; p. 5 © Edgar Fahs Smith Collection, University of Pennsylvania Library; pp. 9, 11, 26, 28 © Lawrence Berkeley National Laboratory; pp. 10, 23 © AP Photos; p. 16 Sam Zavieh; p. 17 © Getty Images; p. 19 © Prof. Kenneth Seddon & Dr. Timothy Evans, Queen's University, Belfast/Photo Researchers; p. 20 © Kansas Geological Survey, www.kgs.ku.edu; p. 22 © David Nicholls/Photo Researchers; p. 25 © G. Otto, GSI; pp. 29, 36 © Popperfoto/Getty Images; p. 32 © www.istockphoto.com/Pali Rao; p. 33 © Russ Lapa/Photo Researchers; p. 35 © Bildarchiv Preussuscher Kultbesitz; p. 38 © Nicola Perscheid/ Getty Images; p. 39 © Hulton Archive/Getty Images.

Designer: Tahara Anderson; Photo Researcher: Marty Levick